People have debated about how our Moon was formed. Some thought that it was a spinning object captured by the Earth when it came too close. Others thought that it was once part of the Earth and came from the Pacific Ocean area before rainwater from comets and volcanoes filled up the low ocean areas.

I0428163

Because our Moon rocks are very much like those found on the Earth, some think the Moon was born when a drifting planet or large rock crashed into the early Earth. Large pieces of Earth were sent up and finally came together to form our Moon.

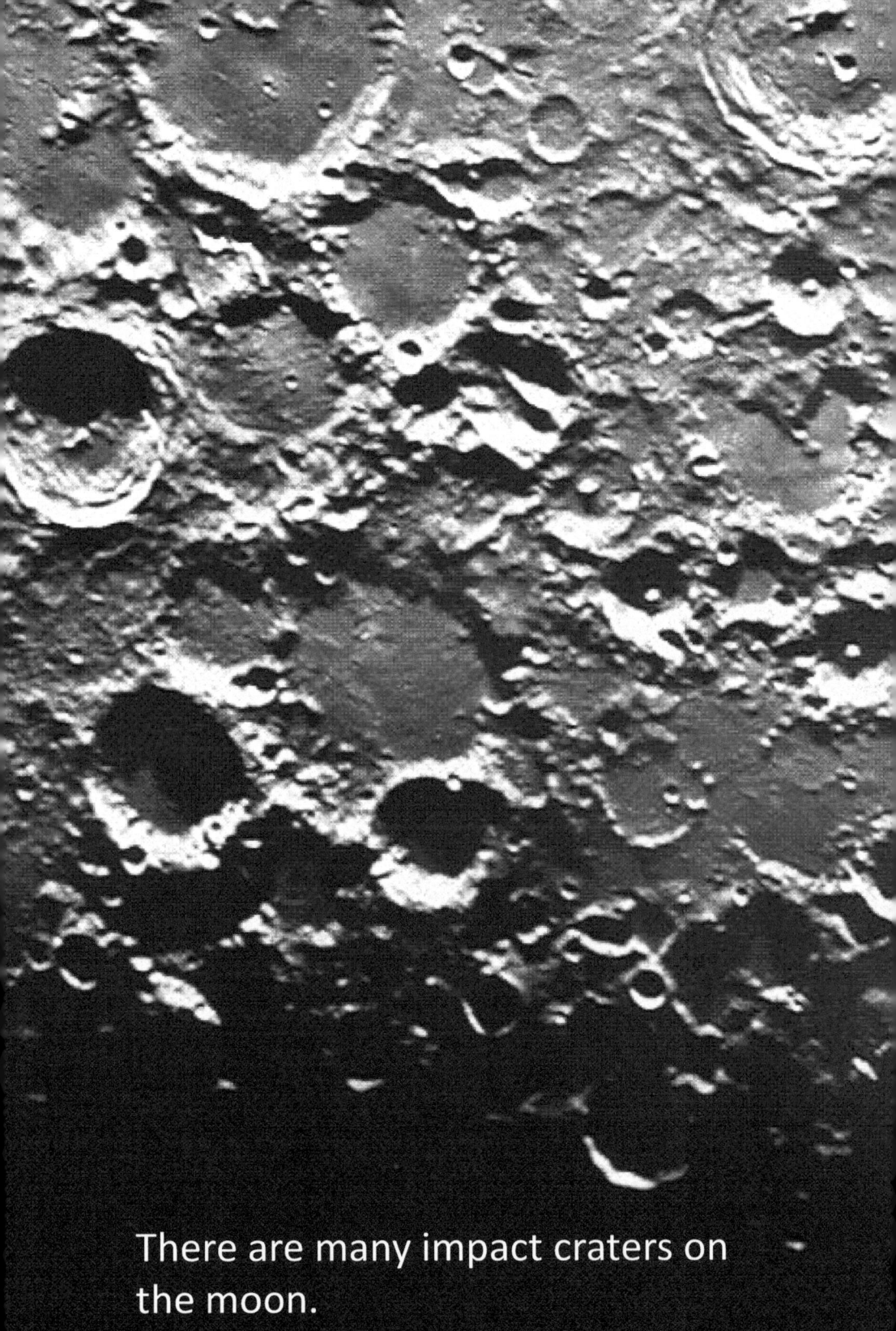

There are many impact craters on the moon.

These impact craters formed when hunks of rocks (meteorites and asteroids), plus rocky icebergs (comets) struck the moon's surface at high speeds (about 30,000 miles per hour or 50,000 kilometers per hour).

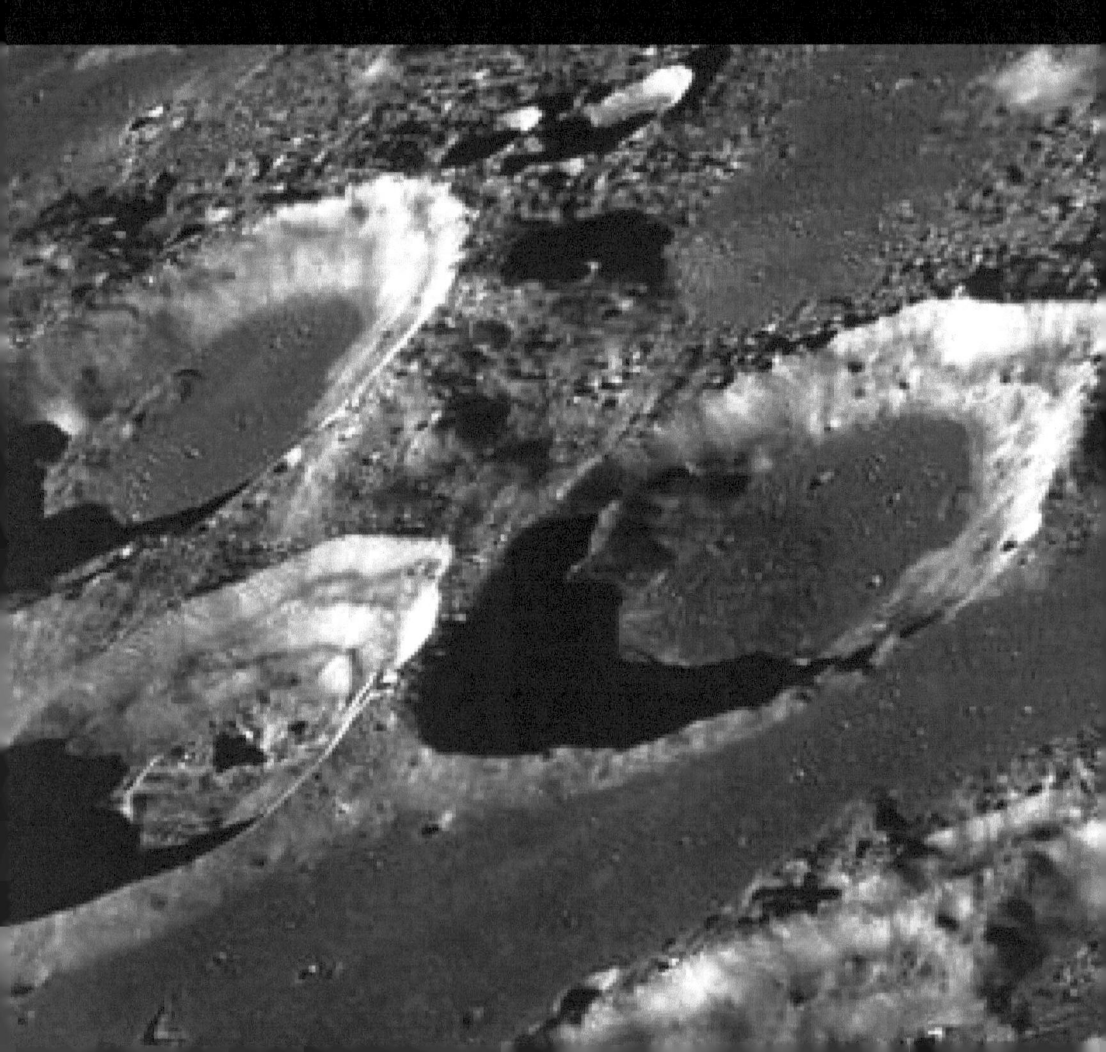

The dark areas on the surface of the Moon were called Seas (Maria) because ancient star gazers thought they were filled with water. They are now known to be of hardened lava called basalt that erupted from ancient volcanoes and fissures.

Does the Moon have active volcanoes like the Earth?

Scientists studied the volcanic rocks brought back from the Moon. They found that long ago the moon had active volcanoes.

The flat dark spots are called
Maria or Seas. They seem to be
where lava flowed into low areas
from ancient volcanic eruptions.
There are no active volcanoes on
the Moon now.

Our Moon could just fit on top of the United States. It is about 2,000 miles (3,000 kilometers) across. Because the Moon is smaller than the Earth, there is less gravity on the Moon than the Earth. How much do you think you would weigh on the moon?

Divide your weight by 6 to find your weight on the surface of the Moon. A dog that weighs 60 pounds on Earth would weigh 10 pounds on the Moon.

60 pounds

10 pounds

Over 2,000 years ago, the Greeks measured the distance to the Moon by comparing it to the size of the Earth. In 1962 a laser beam on Earth was bounced off the Moon to measure the distance. About 50 earths can fit between the Moon and Earth.

In 1969, reflectors were placed on the Moon by astronauts for more accurate measurements. The latest measurement between the Earth and the Moon averages about 200,000 miles or about 400,000 kilometers. It takes about one and a half seconds for Moon light to reach the Earth?

President Eisenhower established in 1958 the National Aeronautics and Space Administration (NASA) for the peaceful use of Outer Space. The Apollo Program (Project Apollo) was created to put 3 humans in space. In 1961, President Kennedy said Apollo's goal was to send humans to the Moon and back safely.

NASA has had over 100 spaceflights since 1958.
Apollo 1 through 10 was practice for Apollo 11 (Apollo Eleven).

In 1969, The Apollo 11 program had 3 Americans who were trained to fly to the Moon. Two would land on and walk on the Moon (Armstrong and Aldrin) and one would orbit around the Moon (Collins).

Michael Collins

Neil Armstrong

Buzz Aldrin

On July 16, 1969, three Americans sat inside the top of a Saturn V (Saturn Five) Rocket. After countdown, they headed to the Moon at speeds of up to 25,000 miles per hour (40,000 kph) to escape Earth's gravity.

On July 20,1969, the three astronauts reached the Moon. While Collins orbited overhead in the Command Module, Armstrong and Aldrin landed on the Moon's surface with the Lunar Module. Neil Armstrong was the first human to set foot on the Moon. His words were, "That's one small step for man, one giant leap for mankind".

Armstrong and Aldrin collected samples of lunar material and set up scientific experiments that transmitted data about the moon's environment.

Because the Moon is so small, the moon's gravity cannot hold air like the Earth can. With no air in outer space or on the Moon, the astronauts had to wear spacesuits that provided them with air to breathe. The carbon dioxide that they breathed out had to be absorbed by a chemical powder. Armstrong and Aldrin gathered samples of lunar material and deployed scientific experiments that transmitted data about the lunar environment.

Because the Moon has very little gravity, there's no real air on the Moon. Astronauts who walked on the Moon had a backpack that weighed about 300 pounds (1,000 newtons) but on the Moon it was about 50 pounds (200 newtons). The backpack had air to breathe like scuba divers on Earth. The American flag was held up by a metal rod across the top.

and 1/2 days later, the Earth gets about 1/4 of the way between the Sun and the Moon, it is called a Crescent Moon.

Full Moon

Crescent Moon

If the Earth gets about 1/2 of the way between the Sun and the Moon, it's called a Half Moon. If the Earth get fully in the way between the Sun and the Moon, it's a New Moon with no sun shining on it.

Half Moon

New Moon

About every 3 and 1/2 days, the Moon changes phases. Starting with the New Moon with no sunlight, waxing means the Moon will shine bigger and more. Waning means the Moon will shine less and smaller.

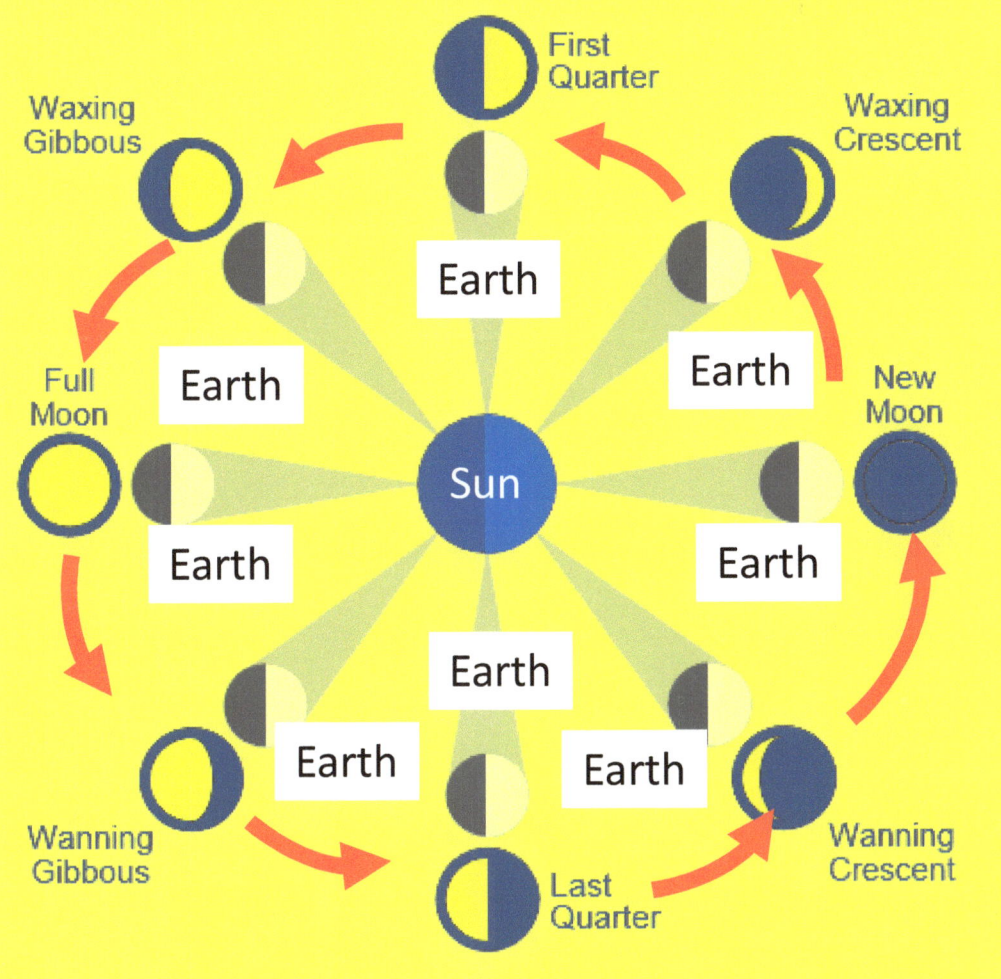

The Eight Phases of the Moon in 29 1/2 days, the phases repeat.
#1. NEW (No sunlight on Moon)
#2. WAXING CRESCENT (sliver of light on right side of Moon)
#3. WAXING HALF or 1st quarter (right half of Moon shines)
#4. WAXING GIBBOUS (the right 3/4 of the Moon is bright)
#5. FULL MOON (the whole Moon is shiny bright
#6. WANING GIBBOUS (the left 3/4 of the Moon is bright)
#7. WANING HALF or last quarter (left half of Moon shines)
#8. WANING CRESCENT (sliver of light on left side of Moon)

It takes about 24 hours for the Earth to spin once on its axis. The Moon takes about 29 Earth days to rotate once on its axis. One day on the Moon is 29 Earth days. Daytime on the Moon is equal to 14 1/2 days on Earth. Nighttime on the Moon is 14 1/2 days on Earth. It takes about 27 days for the Moon to go around or orbit the Earth. Once a year a Super Moon appears 14% larger when the moon's orbit is close to Earth.

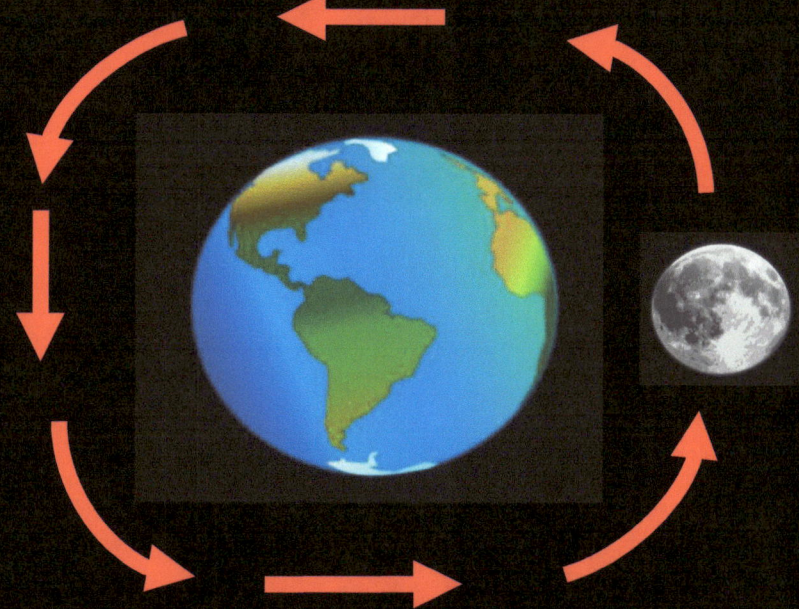

The Moon has no air to hold in heat at night. Without a spacesuit you would freeze solid. The moon's surface can get as cold as minus 279 Fahrenheit (minus 173 degrees Celsius.) During the daytime with no air, the surface of the Moon is about 212 Fahrenheit (100 degrees Celsius.) Your blood would boil without a spacesuit.

The moon's gravity pulls on the ocean to create high tides in line with the moon and low tides when the ocean is not in line with the moon. During high tide, the ocean will cover most of the beach. During low tide, the ocean moves away from the beach,

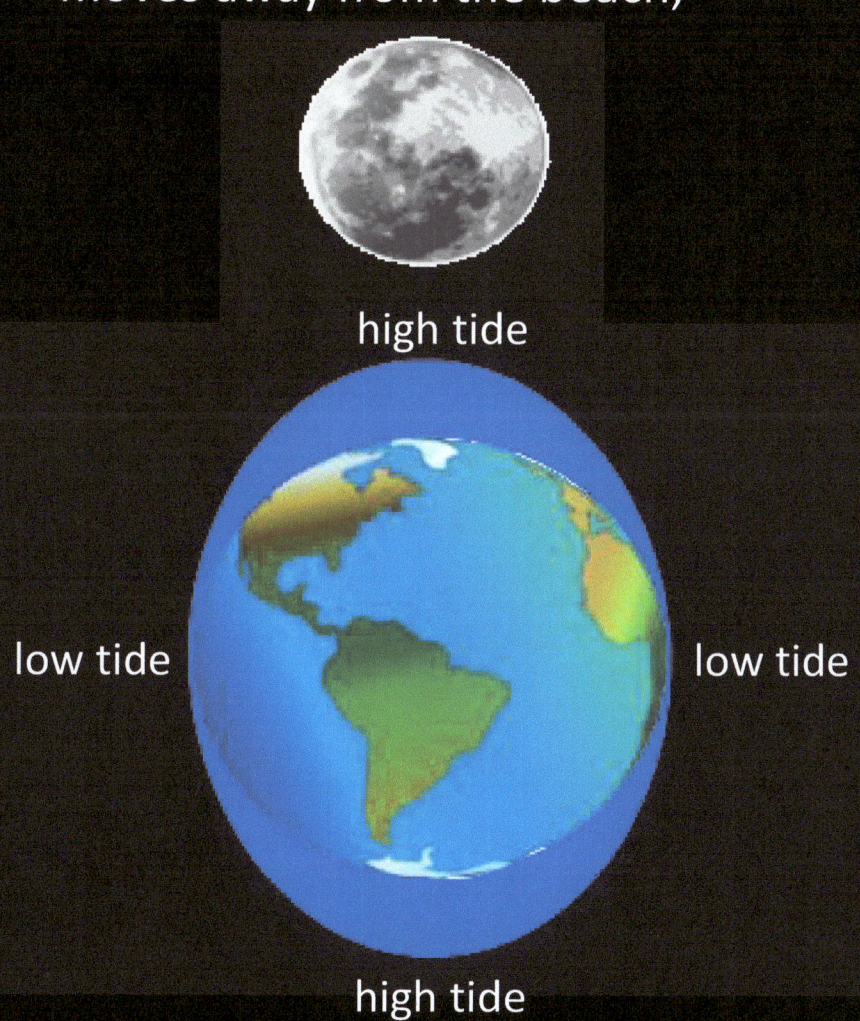

high tide

low tide   low tide

high tide

Maybe, someday you'll visit the moon.
   Dedicated to my lovely wife Sulastri and my grandchildren Mia and Kai as well as those who enjoy learning about the Moon.
   Please follow my author page for new updates at Amazon.com/author/richlinville
   Illustrations from OpenClipArt, PixaBay, Commons Wiki, NASA, and purchased from Edu-Clips

Our Sun

and Solar Eclipses

By Rich Linville

Wolves

by Rich Linville

Sharks

by Rich Linville

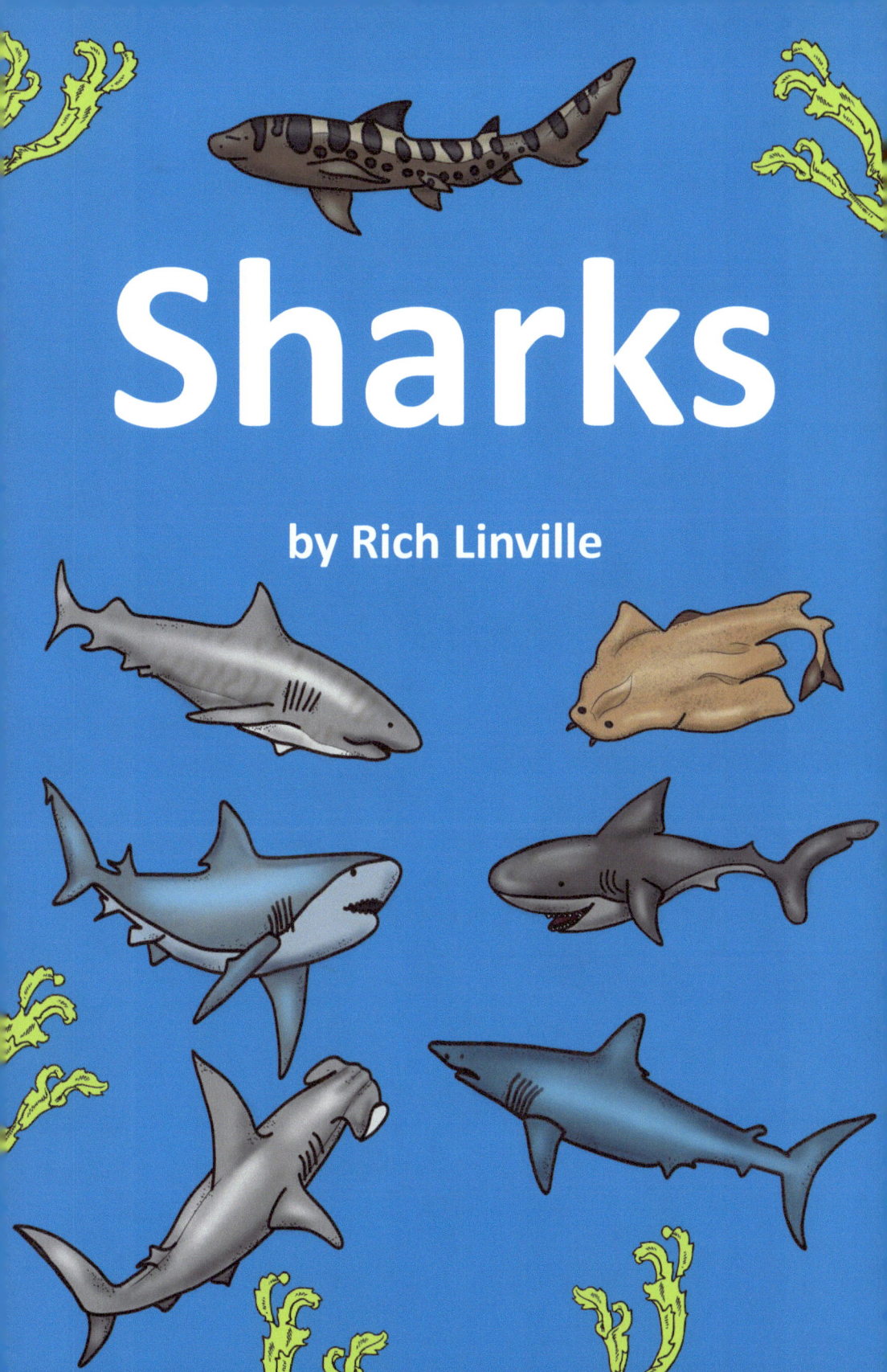

Endangered Animals

by Rich Linville

BATS

and their Life Cycle

by Rich Linville

ISBN: 9798858157700